Introduction

Mathematics can be fun and beautiful. And one can even make money with it by betting. Here are a few examples and lots of mathematical curiosities, collected over many years.

- money winning bets

- the Monty Hall show with a car and 2 goats behind 3 doors

- cyclic winning à la Martin Gardner

- an impossible equation

- identical birthdays; the start of a Formula 1 car race

- Tattoo Formations

- the mystery of the fairy tales from 1'001 Arabian nights

- Euler beauties

- squaring, the easy way, "mathematical beauties"

- the magic number 70 for exponential growth

- the Last Judgment in 800 years?

- birthday and mortality rates, world wide and locally

- the choice from 1'000 balls determining your birth place

- Friday the 13th

- the power of $E=mc^2$

- a collection of riddles

Werner Joho was born 1938 in Baden, Switzerland, and now lives with his wife Rosa in Wuerenlingen, near Zurich. Like Einstein, he attended the cantonal high school in Aarau, and went on to study physics at ETH, the Swiss Federal Institute of Technology in Zurich. There he met Prof. Eduard Stiefel, an outstanding teacher, from whom he acquired a life-long enthusiasm for "Applied Mathematics".

Then followed a fortunate tourning-point when he joined the cyclotron group of Prof. J.P. Blaser, which was then working on a project for a new high intensity proton cyclotron. Calculating the orbits in such an accelerator became his main duty, initially in 1962 at CERN, the European Research Center in Geneva. Later he helped to optimize the accelerator facilities, like the Swiss Light Source, at the Swiss Research Institute SIN/PSI in Villigen. In 1970 he graduated Ph.D. from ETH Zurich.

As a hobby, the author and two ETH colleagues developed a computer chess program. On October 7th 1968 this program played for the first time via ham radio a live game over the Atlantic with a computer program at MIT in Boston. The American program won after 41 moves. The moves of this historic game can be found e.g. with Google: "chess programming Joho".

Over the years the author has collected many mathematical curiosities, a selection of which are published in this booklet. Many chapters should be of interest not only to specialists but to general readers as well.

In his free time Werner Joho has enjoyed many sports activities, like orienteering competitions in the forests, skiing, tennis and windsurfing. For the past 30 years he has been an enthusiastic golfer.

2

The joy of mathematics

Mathematics, …not for you? That's a pity, because you can have lots of fun with it. E.g. you can tempt your friends into some puzzling bets. Then there exist mathematical topics and riddles that stimulate your brain. Most interesting are those riddles which look difficult, but are easy to solve with a simple trick.

Did you know: That is easy to estimate the number of births and deaths per year worldwide or in your home town?

That at a party with 40 people there is a 90% chance, that two have a common birthday?

That you can test the multiplication skill of your children without doing the multiplying yourself?

That you are privileged, because you could have had a 50% chance to be born in a neighbourhood without sanitary installations, and with a future income of less than 3$ a day?

Dear reader, I wish you lots of entertainment and pleasure from reading this booklet. I encourage you to use or cite examples from it. After all, we want to make the application of mathematics in our daily life more popular, don't we?!

Werner Joho

Wuerenlingen, 25.2.2018

1.edition

Production and Publishing House:

BoD – Books on Demand, Norderstedt, Germany

ISBN 978-3-7460-5623-4

Originally published in 2017 in German entitled
"Wetten gewinnen dank Mathematik"
under ISBN 978-3-7448-1113-2
Many thanks to John Crawford, who helped
with the translation from German to English

Multiplication of Probabilities

In order to understand better the calculation of probabilities for the following bets, I present here a simple example. What is the probability p, that one throws with a die first an even number, and in the second throw a 6?

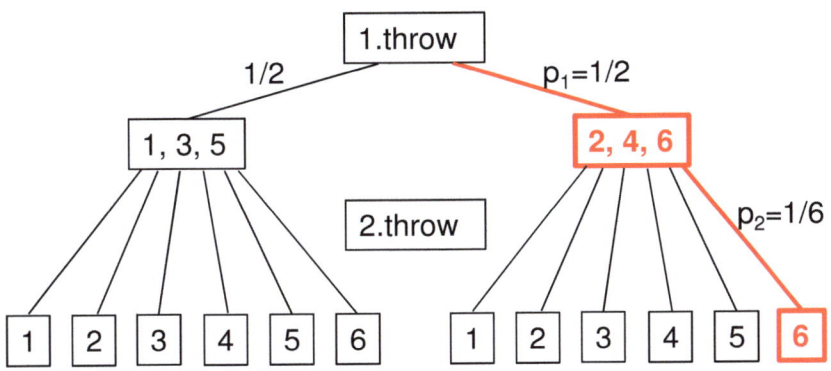

For this probability p one has to **multiply together** the successive probabilities p_1 und p_2 for the 1st und 2nd step. Because only one out of the 12 possible outcomes is the desired one.

$$p = p_1 \cdot p_2 = \frac{1}{2} \cdot \frac{1}{6} = \frac{1}{12}$$

Throw 6 dice onto a table; this costs you 1 Swiss frank as a bet.

If all 6 numbers (1, 2, 3, 4, 5, 6) show up once each,

I pay you **30 franks** ;

is this a fair bet?

No! **The chance p of winning is only 1.5% !!**

To be fair, I should offer you 64 franks !

$$p = \frac{5}{6} \cdot \frac{4}{6} \cdot \frac{3}{6} \cdot \frac{2}{6} \cdot \frac{1}{6} = \frac{5!}{6^5} = \frac{5}{324} = 1.5\%$$

Chance = 9% for 5 different numbers after 5 throws (the first throw can give any number)

For this calculation the probabilities for each successive throw have to be multiplied together. If the first 5 numbers are different (the chance is only 9%), then the chance to get in the last throw the missing number is only 1/6, which amounts to 1.5% for the overall probability.

A game for wedding parties

I often perform this betting game with 6 dice at weddings. Everybody is encouraged to throw the dice many times, for one frank a throw. This income goes to the newly-wedded couple!

My contribution to the couple is to pay for any wins, out of my own pocket! Here are the chances, that I have to pay n times the 30 franks (= success for the thrower) with 100 throws, according to so-called "Poisson statistics".

My average contribution is about 100x30franks/64 = 47 franks. With a fair payment of 64 franks my average contribution would be 100 franks, the same amount as the income from 100 throws. But consider that the chance, that I have to pay 5 times or more the 30 franks is about 2%.

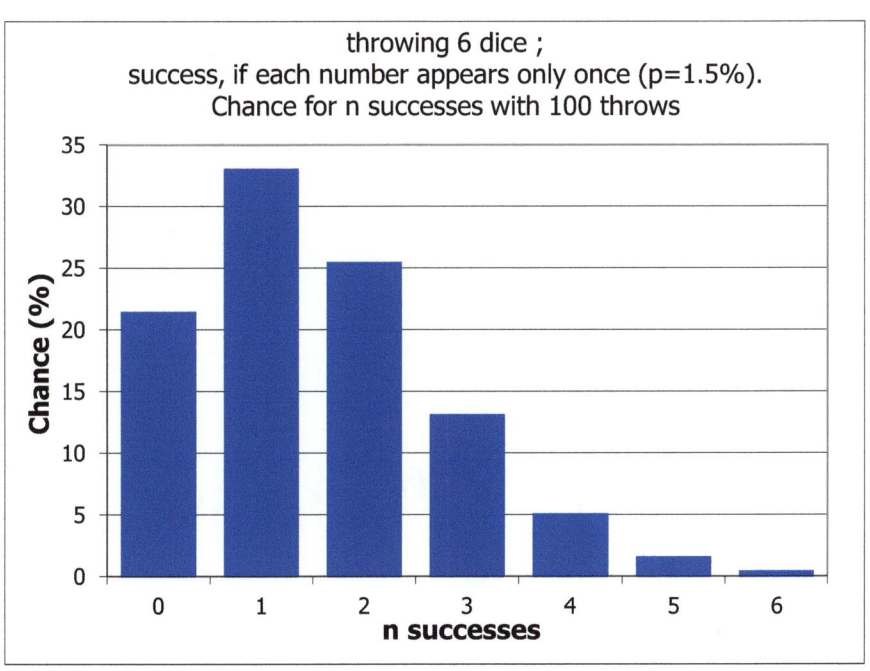

throwing 6 dice ;
success, if each number appears only once (p=1.5%).
Chance for n successes with 100 throws

What is the chance p to throw with a die at least one 6 in 3 trials?

The chance for a 6 in the first throw is 1/6.

But the chance for at least a 6 in 3 trials is not

3 x 1/6 =1/2. It is less:

$$P = 1 - (\frac{5}{6})^3 = \frac{91}{216} = 42\%$$

To make this plausible: after 6 trials the chance is not 6 x 1/6 = 100%. It is only

$$P = 1 - (\frac{5}{6})^6 = 66.5\%$$

Another unfair betting offer

Draw four cards from a bridge deck (52 cards).

If you get exactly one card from each suit (heart, diamond, spade, club), I pay you **five times** the sum you bet.

Is this fair?

No! **Your winning chance p is only 10.5%** !!

$$p = \frac{39}{51} \cdot \frac{26}{50} \cdot \frac{13}{49} = 0.105$$

Chance = **40%** for 3 different "suits" after 3 cards (the suit of the first card is irrelevant).

From the remaining 49 cards only 13 are from the missing "suit"; this means a chance of **26%**. The overall chance is thus 40% of 26% = **10.5%**.

A common card from 2 bridge decks

You take from a deck of 52 bridge cards a subset
of 7 cards. I do the same from another full deck.
I win the bet if we have at least one common card.
What is **your** chance of winning? => Only 1/3!

$$p_7 = \frac{45}{52} \cdot \frac{44}{51} \cdot \frac{43}{50} \cdot \frac{42}{49} \cdot \frac{41}{48} \cdot \frac{40}{47} \cdot \frac{39}{46} = 34\%$$

If I draw **six** cards, but you can draw **seven**
cards, your winning chances are increased to:

$$p_{67} = \frac{45}{52} \cdot \frac{44}{51} \cdot \frac{43}{50} \cdot \frac{42}{49} \cdot \frac{41}{48} \cdot \frac{40}{47} = 40\%$$

Only with **six** cards for each of us is your chance
of winning almost fair:

$$p_6 = \frac{46}{52} \cdot \frac{45}{51} \cdot \frac{44}{50} \cdot \frac{43}{49} \cdot \frac{42}{48} \cdot \frac{41}{47} = 46\%$$

Quiz with Three Questions

In this quiz (called "Time is Money"), which is often presented on a Swiss radio station, one has to answer three questions about a total of nine events. In each question one has to put given events in the right order in time.

In the first question one has to decide which of two events happened first. In the second, we have to place three events in order, such as

1. Hillary and Sherpa Tensing climb Mount Everest.

2. Germany wins the soccer world championship for the first time.

3. Outbreak of the war in Korea

In the third and final task one has to classify four events.

Let's suppose that we have no idea when all these nine events happened. What is then the chance to answer all three questions correctly, simply by guessing?

For the first question the probability is 1/2, for the second one it is $1/3 \cdot 1/2 = 1/6$ and for the third one only $1/4 \cdot 1/3 \cdot 1/2 = 1/24$. Multiplied together the final probability is only 1/288 = 0.35% to answer all three questions correctly! This explains why only a few candidates win at this game.

Choose from the 25 letters of the alphabet (I and J are the same) **5 different letters** and write them on a sheet of paper, which you hide from me.

I do exactly the same.

Then we compare our letters.

If we have <u>no</u> common letter, **you will win the bet,** otherwise I win it.

How big is the chance p_5 that you win?

$$p_5 = \frac{20}{25} \cdot \frac{19}{24} \cdot \frac{18}{23} \cdot \frac{17}{22} \cdot \frac{16}{21} = 29\%$$

You can play simultaneously against a whole group!

This increases your chances of winning.

In the first round you can increase your chance for

winning slightly due to a psychological effect:

Some opponents tend to choose letters like (Q, X, Y..!)

This game is only somewhat fair if both parties choose
4 letters, the chance for the opponent to win is then:

$$p_4 = \frac{21}{25} \cdot \frac{20}{24} \cdot \frac{19}{23} \cdot \frac{18}{22} = 47\%$$

The Monty Hall show with 3 doors

This game is based on an idea of the statistician Steve Selvin (1975). It became famous through the **Monty Hall TV show.** It caused a huge controversy after a newspaper column by Marilyn vos Savants in 1990. In letters to the editor many established mathematicians disgraced themselves by claiming that Marilyn was talking bullshit!

In his show Monty Hall presented three doors to a candidate. Goats were hidden behind two doors, while behind the third was a car, its position known only to Monty Hall. The candidate was now given the **chance to win this car**, if he pointed to the corresponding door. The chance to win the car was thus **1/3**. Lets assume the candidate chose door number 1. From the remaining two doors Monty Hall would now open a door, behind which he knew a goat was placed (e.g.door 3). The car was now for sure behind door 1 or 2. Then came the showdown: the candidate was given the **chance to switch** from his original choice (door 1) to door 2.

Now the big question: is it irrelevant, if the candidate switches or not, or should he always switch?

Always switch doors!

=> The chance to win the car increases to 2/3!!

Without switching the chance remains 1/3 (and is not 50% as many people expect).

Even "experts" are surprised by this. There exist many explanations for the result. I offer here a simple experiment, which should be understandable even without mathematics.

Lets imagine that I play 30 times again the show master.

He agrees to hide the car randomly, 10 times each behind the doors 1, 2 and 3. My strategy is very simple: I always choose door 1, and then switch doors! Then I leave the stage to drink a coffee. After the 30 trials are over the show master can call me back.

How does the final result look?

Remember that the show master never opened my chosen door 1.

- 10 times the car was behind my chosen door 1.

Then door 2 or 3 was opened. I switched doors…and lost.

- 10 times the car was behind door 2. The show master **had** to open door 3.

I switched to the closed door 2…and won.

- 10 times the car was behind door 3. The show master **had** to open door 2.

I switched to the closed door 3…and won.

=> I won 20 times and lost 10 times.

My chance of winning: **2/3 as predicted!!**

Without switching I would have had only 10 wins against 20 losses.

Published by Martin Gardner in "Scientific American" in Dec. 1970
(110-114): There are 4 specially prepared dice, each showing a
different arrangement of points from 0 to 6. I let you choose first
one of these dice. Then I make my choice, depending on your
choice! Then we throw our dice. The rule is very simple: The bet
is one frank each. The higher point showing wins the bet.

In the long run my chances of winning are 2:1, no matter which
die you choose **first**!

Here are the 4 dice:

Die	points showing	point average
A	4 4 4 4 0 0	2.67
B	3 3 3 3 3 3	3.0 (no need to throw it!)
C	6 6 2 2 2 2	3.33
D	5 5 5 1 1 1	3.0

Die A beats Die B   with 24 wins and 12 losses
Die B beats Die C   with 24 wins and 12 losses
Die C beats Die D   with 24 wins and 12 losses
Die D beats Die A   with 24 wins and 12 losses

The interesting point is, that we have a
cyclic winning pattern: There is no best die!

Cyclic winning for wrestling teams

I modified the Martin Gardner puzzle from 4 dice to 4 wrestling teams. In each team there are 6 wrestlers. Each wrestler has a rating for his strength; the stronger wrestler has a higher rating and always wins his match against a weaker wrestler. In a team competition every wrestler fights against every opponent from the other team. There are thus 36 matches.

The following 4 teams take part in the competitions:

Team	individual strengths						average strength
A	17	17	17	17	6	6	13.33
B	13	13	13	13	13	13	13.0
C	20	20	8	8	8	8	12.0
D	18	18	18	7	7	7	12.5

Team A beats team B   with 24 wins and 12 losses

Team B beats team C   with 24 wins and 12 losses

Team C beats team D   with 24 wins and 12 losses

Team D beats team A   with 24 wins and 12 losses

Again we have a cyclic winning pattern! Each team is beaten by one specific team.

illustrated in a graph:

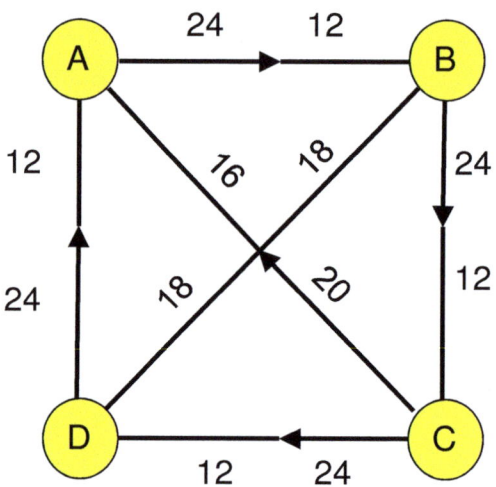

In a **round robin** competition each team plays the other three teams. Each team gets 2 points for a win, and 1 point for a draw.

The final score will look like this:

Result	Team	wins	losses	draws	points	team strength
1.	C	2	1	0	**4**	12.0
2a.	D	1	1	1	**3**	12.5
2b.	B	1	1	1	**3**	13.0
4.	A	1	2	0	**2**	13.3

The result in a round robin is thus just opposite to the average team strength:

the "weakest" team wins, the "strongest" team loses!

With a play-off or **cup system** we have 3 cases:

Case 1.   Semifinals:

A – B ,   A wins 24:12;   C – D ,   C wins 24:12

Final: A – C ,   **C wins** 20:16

Case 2.   Semifinals:

A – C ,   C wins 20:16;   B – D , draw 18:18,

                         D wins Tie break between the 2 top wrestlers

Final: C – D ,   **C wins** 24:12

Case 3. Semifinals:

A – D ,   D wins 24:12;   B – C , B wins 24:12

Final: B – D , draw 18:18,

D wins Tie break between the 2 top wrestlers

Again: the "weakest" team C (lowest average strength) wins 2 of the 3 possible playoffs !

It is important, in what sequence the teams start the semifinals. Team D has a chance to win only if it can start against Team A. A and B, the two "strongest" teams, can never win the cup!

Such a cyclic winner scheme probably decided the **presidential election of 2016 in the US**: Donald Trump won over Hillary Clinton, Clinton won over Bernie Sanders. But Sanders probably would have beaten Trump!

The selection of players among youngsters

Fourteen boys get together to play soccer.

The captains of Team A and Team B can each select 6 players.

Their strength is ranked from 1 to 12.

If Teams A and B **alternate** with their choice of players, then this is unfair for team B!

Team A obtains the "uneven" players 1,3,5,7,9,11 with 36 as the sum of their rankings. Team B obtains the "even" players with 42 as the corresponding sum.

Correct is:

Captain A chooses player 1

then Captain B chooses players 2 and 3

then Captain A chooses players 4 and 5

then Captain B chooses players 6 and 7

then Captain A chooses players 8 und 9

then Captain B chooses players 10 and 11

finally Captain A gets player 12

Team A: player 1, 4, 5, 8, 9, 12;   sum of rankings = 39

Team B: player 2, 3, 6, 7, 10, 11; sum of rankings = 39

Team A gets the best, as well as the lowest ranked player.

The impossible equation !?

$$X - 3 + 2 = X$$

X = 5 matches

1. Remove 3 matches

2. then add 2 matches to get the same picture !?

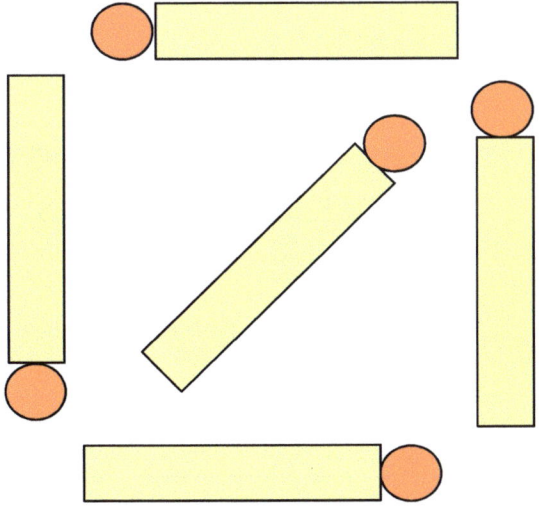

warning: this problem can drive people crazy!

(think a while, before you see solution on next page)

1. remove 3 matches

2. add 2 matches

to get the same picture !

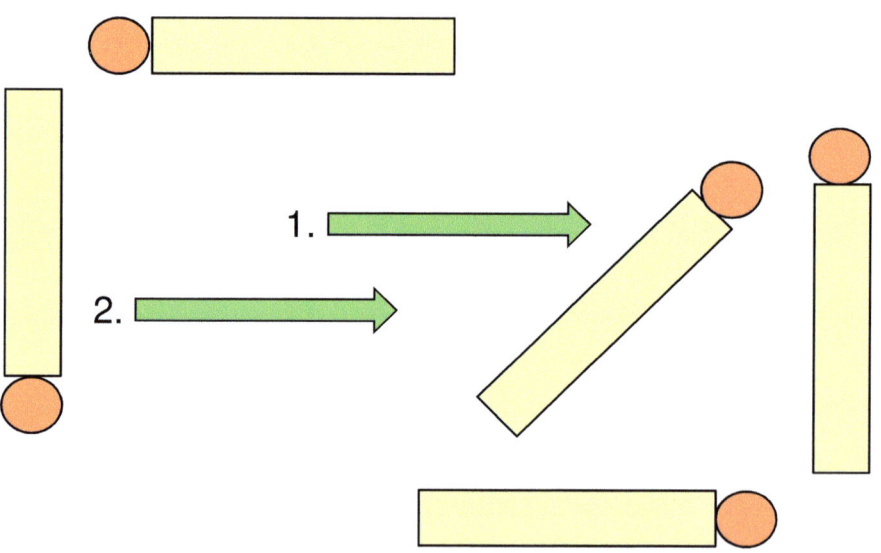

Bet for a common birthday in a group

What is the chance that in a group of n persons, two people have the same birthday (disregarding the 29th February)?

$$\text{Probability } p = 1 - \frac{364}{365} \cdot \frac{363}{365} \cdot \frac{362}{365} \cdot \ldots \cdot \frac{(366-n)}{365}$$

With **23 people** the chance is already 50% and with 50 people even 97%. The chance to have two identical birthdays or one day apart is with 20 people already 80%. This gives a very good bet!

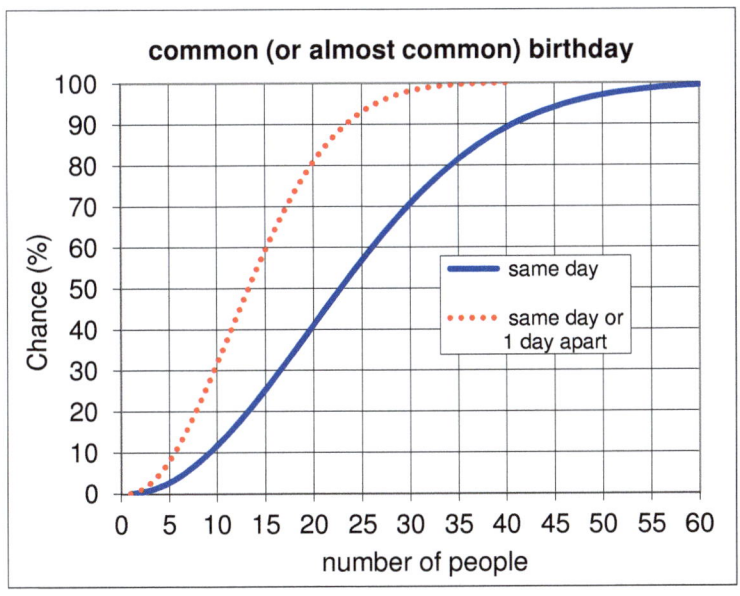

Quadratic birthday gifts for two brothers

Walter and John are brothers. Walter is 13 years old. His younger brother John is 12 years old, but the brighter one of the two. John proposes to his brother the following scheme for their birthday gifts:

Let's give each other **gifts**, which increase **quadratically** in value with each year. We start this year with a gift of 1 frank. Next year the gift has a value of 4 franks, the following year it is worth 9 franks and so on. Why don't you start the series this year with 1 frank. Since I am one year younger than you, I will start with my series next year with 1 frank. Is this OK with you? Walter agrees, and does not realize that each year the **difference** between their gifts **increases linearly for the rest of their lives**!

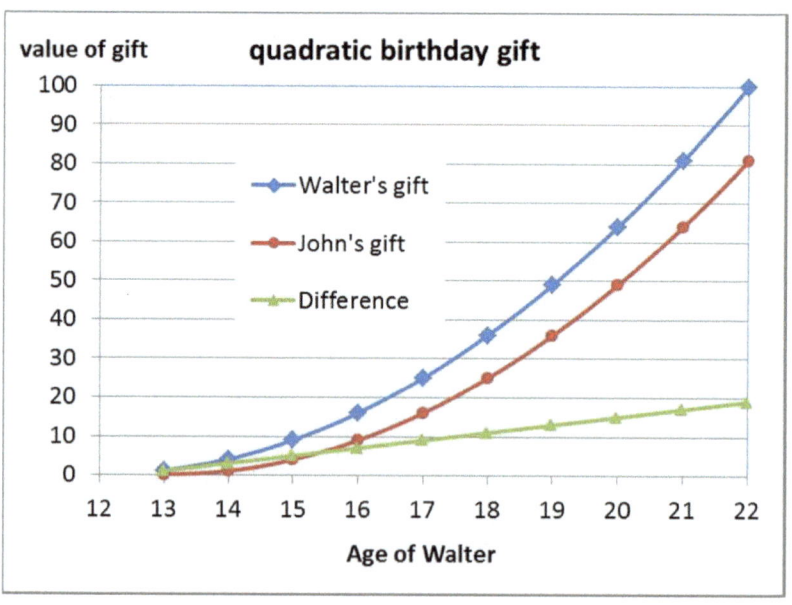

A situation, similar to the quadratic birthday present, happens at the start of a car race. The cars can accelerate to 200km/h (55m/s) within 5 seconds. This amounts to an acceleration of 11m/s^2, about 10% more than the acceleration due to gravity. At the start a driver reacts 0.5 s later than his opponent next to him. Right at the start he is already 1.4m behind. And it gets worse: With the same acceleration the distance to his opponent increases linearly with time. **After 5 seconds** he is still 0.5s behind, but at a velocity of 55m/s this amounts to a **difference of 27m =** 0.5sx55m/s. He covered 112m against the 139m of his opponent.

Formula 1

Driver 2 starts   0.5 s after Driver 1

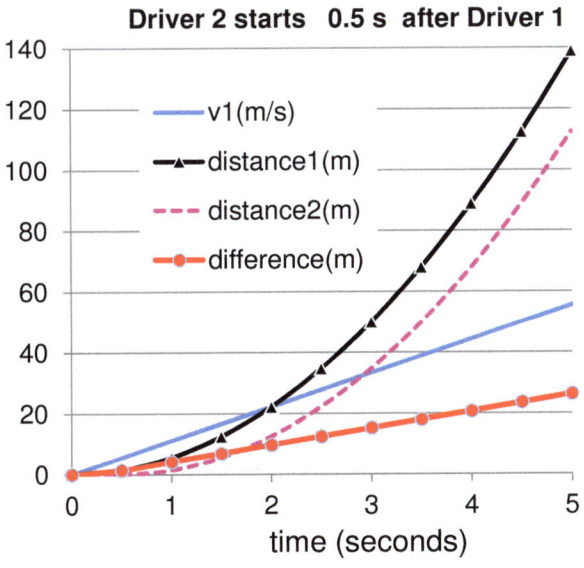

time (seconds)

Crashing cars on a highway

The reverse case to the start of a race is a crash on the highway. Here two cars are travelling behind each other, **32m** apart, both at 120km/h (33.3m/s). Suddenly the front car initiates a full stop, coming to rest after 67m, 4s later. The driver in the back car steps on the brakes as well, but after a reaction time of 1s. His car would come to a full stop after 4s as well. But the distance to the front car decreases continuously, and after 3.5 s the second car crashes with 15km/h into the front car, which is at rest.

The simple rule to avoid a crash: **the distance to the front car has to be bigger than the distance which is covered during the reaction time.**

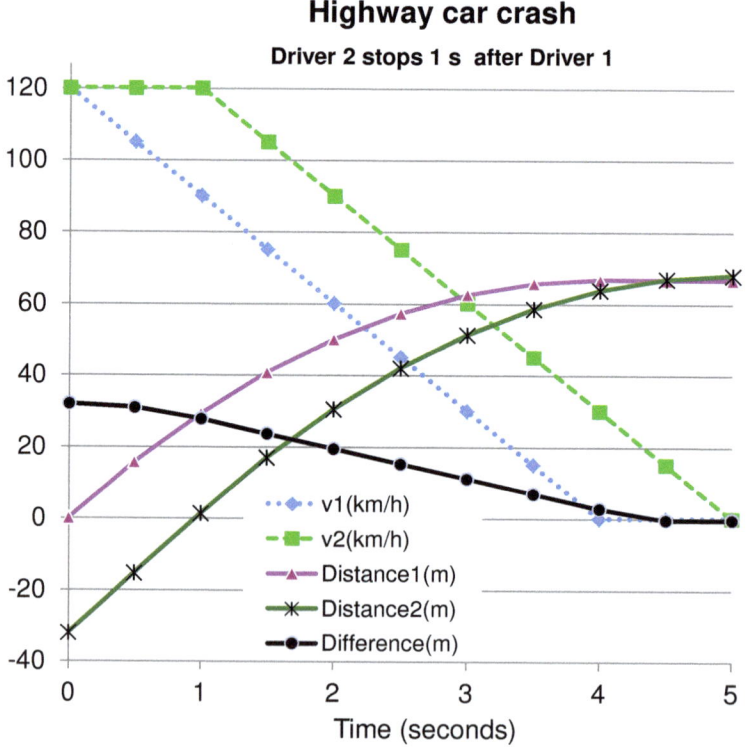

Highway car crash

Driver 2 stops 1 s after Driver 1

Time (seconds)

Friday the 13th

Are you superstitious ?

What is **N**, the number of days you have experienced so far?

Simple estimate:

- each year has a 13th day of the month, 12 times
- the chances for a Friday are 1/7

This means almost two (12/7=1.7) Friday the 13th each year

$\Rightarrow$ **N $\approx$ your age x 12/7**

$\Rightarrow$ at age 58 you have survived a Friday the 13th about 100 times

Harmonic Rectangular Formations

h is, according to the mathematician Ramanujan, a harmonic number, which can be partitioned in **p** different ways into a product **h**=AxB, (BxA counts as a new product).

There are thus **p** possibilities, to form with **h** persons a rectangular formation e.g. for the so-called Tattoo-Festival.

h	partitioning	p Products AxB
2=2!	2	2
4	2x2	3
6=3!	2x3	4
12	2x2x3	6
24=4!	2x2x2x3	8
60	2x2x3x5	12
120=5!	2x2x2x3x5	16
180	2x2x3x3x5	18
240	2x2x2x2x3x5	20
360	2x2x2x3x3x5	24
720=6!	2x2x2x2x3x3x5	30
5040=7!	2x2x2x2x3x3x5x7	60

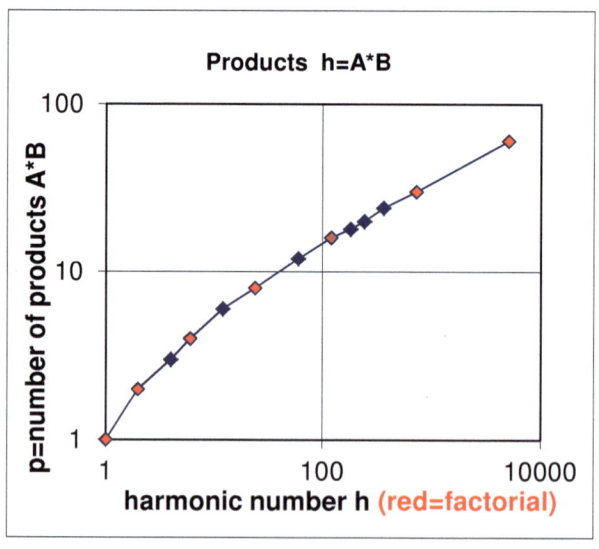

Products h=A*B

p=number of products A*B

harmonic number h (red=factorial)

Tattoo Formations

P=8 rectangular formations with h=24 persons

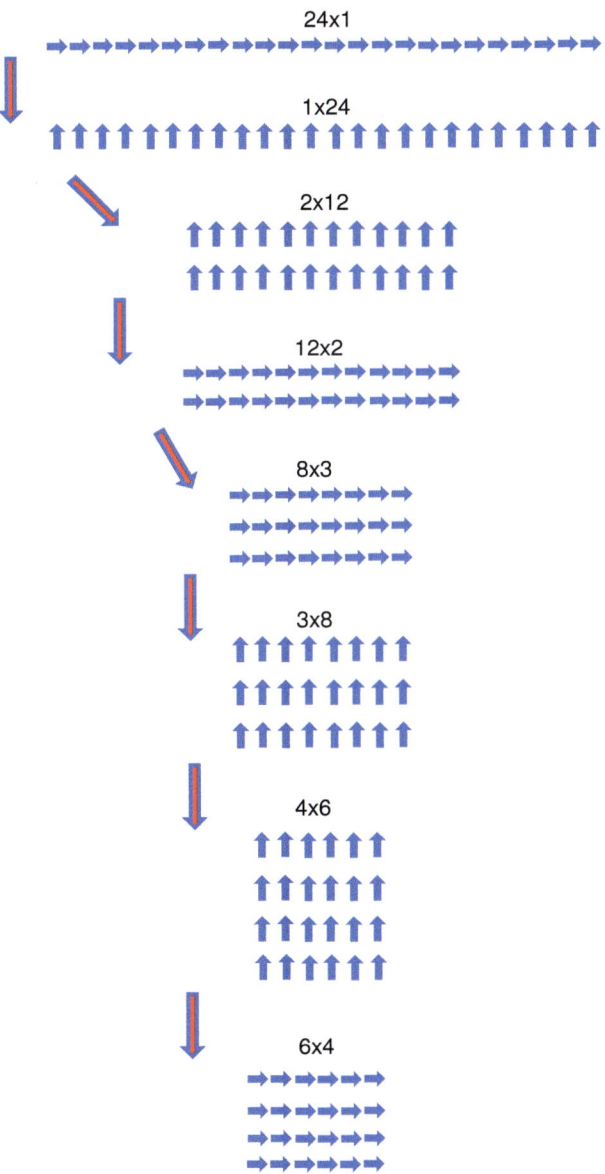

24x1

1x24

2x12

12x2

8x3

3x8

4x6

6x4

Tattoo Formations

P=12 rectangular formations with h=60 people

60x1

1x60

30x2

20x3

2x30

3x20

15x4

12x5

10x6

6x10

4x15

5x12

- Take a 3-digit number **xyz (e.g.597)**

- multiply it by **7**

- then multiply the result by **11**

- and this result finally by **13**

Result :

$7 \cdot 11 \cdot 13 \cdot xyz = $ **1'001** $\cdot xyz = $ **xyz'xyz**

as an example:

$7 \cdot 11 \cdot 13 \cdot$ **597** $= $ **1'001** $\cdot$ **597** $ = $ **597'597**

The Mystery of 1'001 Arabian Nights

The secret behind the 1'001 arabian fairy tales is finally revealed:

Sheherazade, the daughter of the visier, avoided execution by the king Sharyar, for 1'001 nights, with her fairy tales.

But why 1'001? Since **1'001 = 7 x 13 x 11** , this means first **7** days (a week), then **13** weeks (a season). Then Sheherazade found out she was pregnant and thus safe. Finally **11** seasons followed!

These **11** seasons are composed of:

3 seasons: first pregnancy

1 season: recovery

3 seasons: second pregnancy

1 season: recovery

3 seasons: third pregnancy

$$\pi \approx \frac{355}{113}$$

$$\pi = 3.141592_7$$

$$\frac{355}{113} = 3.141592_9$$

This approximation of π with a rational number was known already to ZuChongzhi (429-500), a Chinese Mathematician and Astronomer !

$$e^{\,i\pi} = -1$$

This beautiful formula, from the Swiss mathematician and physicist Leonhard Euler (1707-1783), contains **two fundamental numbers** (e, π) and **three important mathematical inventions:**

1. the equal sign = (replacing the words: "**is equal to**") was invented in 1557 by the Welsh mathematician Robert Recorde who stated:
   "no two things can be more equal than these two bars"

2. the negative numbers

3. the imaginary numbers with the unit i

Euler himself invented the symbols for e and i, as well as the symbols Σ for a sum and f(x) for functions.

The Euler cubes

Here are 6 cubes with side lengths of

1, 2, 3, 4, 5 and 6 units.

Can you build, with the help of cubes like these,

two piles which have the same volume?

The solution, with a little trick, follows from the

Euler formula on the next page.

A nice one from Euler is:

$$3^3 + 4^3 + 5^3 = 6^3$$

The Indian Mathematician Ramanujan found:

$$9^3 + 10^3 = 1^3 + 12^3 = 1'729$$

Disproving a conjecture by Euler, a computer program found in 1966 that:

$$27^5 + 84^5 + 110^5 + 133^5 = 144^5$$

I figured out the nice formula below myself; it is however already mentioned in a book by the Hungarian mathematician George Polya.

$$1^3 + 2^3 + 3^3 + ... + n^3 = (1 + 2 + 3 + ... + n)^2$$

$$\frac{(n+1)^2 n^2}{4} \qquad [\frac{(n+1)n}{2}]^2$$

Example: arrangement of 225 single cubes to a

15x15 square or to cubes of sides 1, 2, 3, 4 and 5

$$1^3 + 2^3 + 3^3 + 4^3 + 5^3 = (1+2+3+4+5)^2 = 225$$

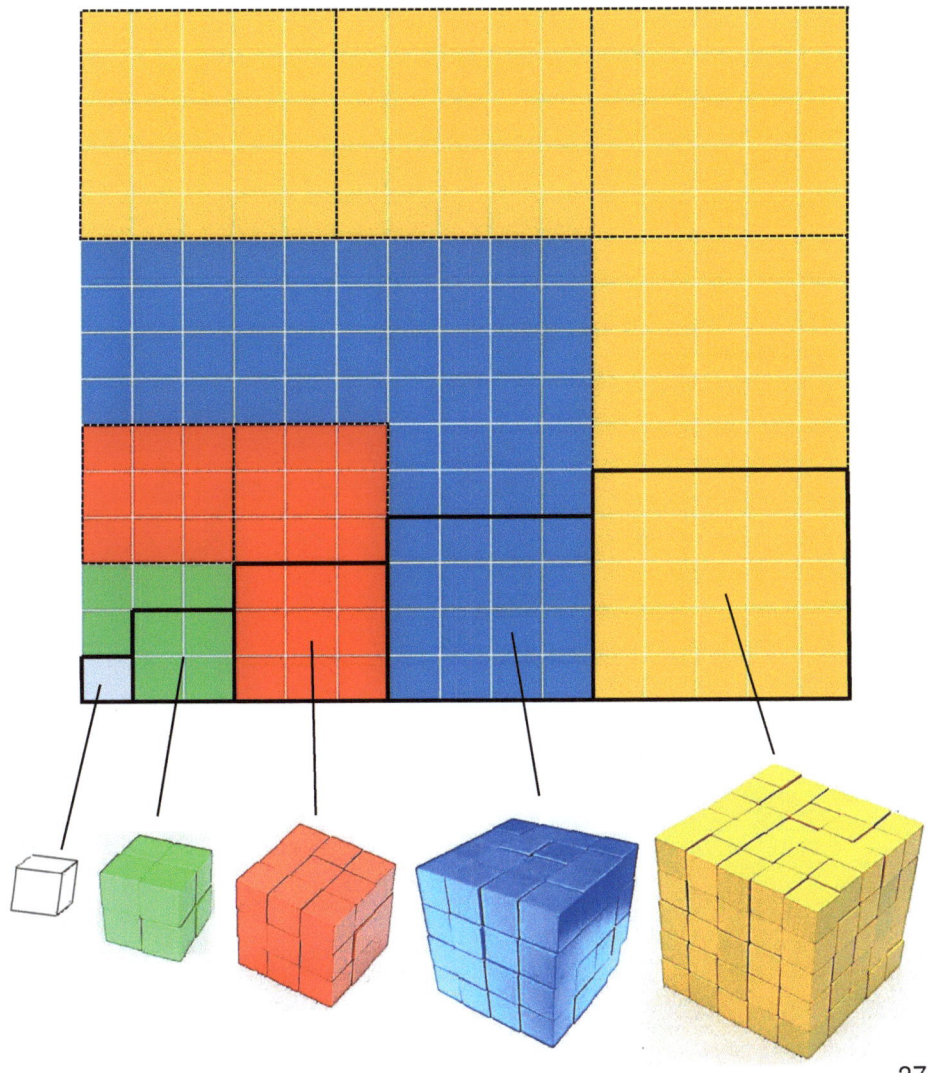

Euler: $3^3 + 4^3 + 5^3 = 6^3 = 216$

rearranged : $7 \times 3^3 = 4^3 + 5^3 = 189$

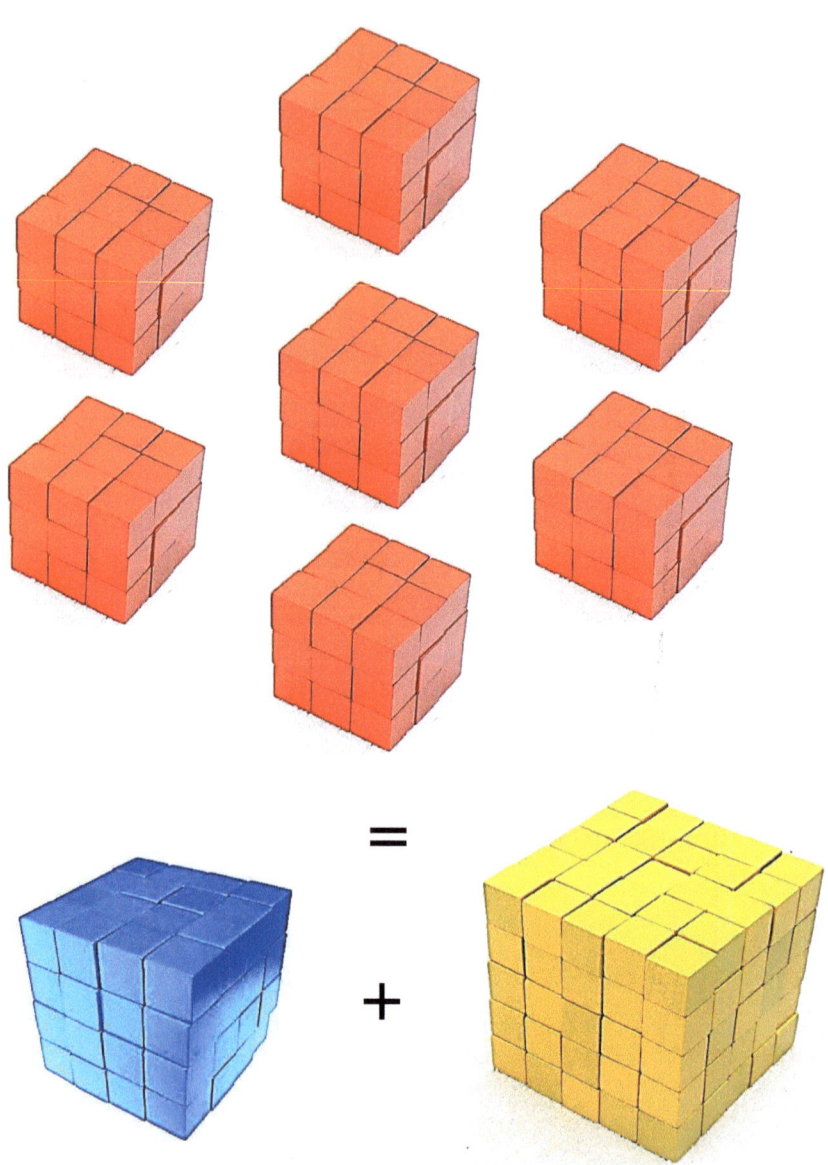

$(a+b)^2$ illustrated graphically

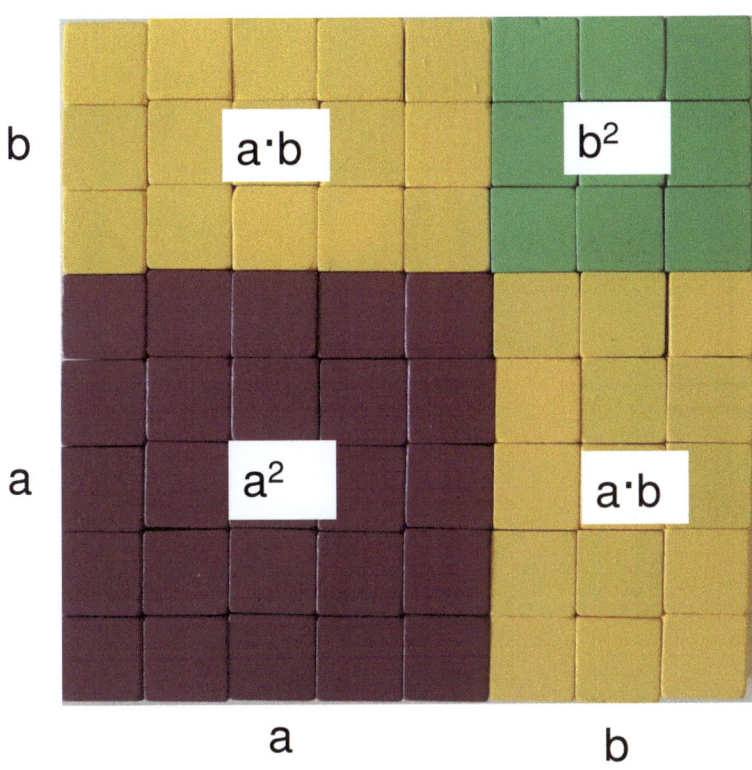

$$(a+b)^2 = a^2 + b^2 + 2 \cdot a \cdot b$$

Example: a=5 , b= 3
$$(5 + 3)^2 = 5 \cdot 5 + 3 \cdot 3 + 2 \cdot 5 \cdot 3$$
$$= 25 + 9 + 30 = 8^2 = 64$$

3.5 x 3.5 in 3.5 seconds

Recipe: $(n-\frac{1}{2})(n+\frac{1}{2}) = n^2 - \frac{1}{4}$

$$n^2 = (n-\frac{1}{2})(n+\frac{1}{2}) + \frac{1}{4}$$

for n = half integer:

n	n^2
2.5	$2 \cdot 3 + 0.25 = 6.25$
3.5	$3 \cdot 4 + 0.25 = 12.25$
4.5	$4 \cdot 5 + 0.25 = 20.25$
5.5	$5 \cdot 6 + 0.25 = 30.25$
6.5	$6 \cdot 7 + 0.25 = 42.25$
7.5	$7 \cdot 8 + 0.25 = 56.25$
...	
11.5	$11 \cdot 12 + 0.25 = 132.25$

Squaring numbers around 50 and 100

1) $x \approx 50$, $x \equiv 50 + n$

$$x^2 = 50^2 + n \cdot 100 + n^2$$

$$x^2 = (x-25) \cdot 100 + n^2 = (25+n) \cdot 100 + n^2$$

e.g. $x = 48$, $n = -2$

$$48^2 = 2'300 + 4 = 2'304$$

this trick is mentioned by the famous

physicists Richard Feynmann and Hans Bethe

2) $x \approx 100$, $x \equiv 100 + n$

$$x^2 = 100^2 + 2n \cdot 100 + n^2$$

e.g. $x = 104$, $n = 4$

$$104^2 = 10'000 + 8 \cdot 100 + 16 = 10'816$$

9 * 123'456'789 + 10 = **1'111'111'111**

888'888'888 : 9 = **98'765'432**

111'111'111 * 111'111'111 =

1234**5678**9**8765**4321

21'649*513'239 = 11'111'111'111

prime numbers

The following formula was constructed

by Leonhard Euler:

$$P(n) = n(n-1)+41$$

It's hard to believe, but for n=1, 2, 3, ... up to 40
this formula gives a **prime number** !
It fails for the first time at n=41, where
P(41)=41*41=1'681
(It then fails further at n=42, 45, 50, 57, 66 etc.)

If you see a series of numbers: 2, 4, 6, 8, 10, 12, ...

created by a formula F(n), for n=1, 2, 3, ...6

you probably guess, that the next term is 14 !?

Now give me the number **Y**, the year you were born.

I give you below a formula F(n), where the next term

in the series (for n=7) is not 14, but exactly **Y** !

F(n)=2n+ (**Y**-14)(n-1)(n-2)(n-3)(n-4)(n-5)(n-6)/6!

The lens equation for the image of an object
with a thin lens of focal length f

$$\frac{1}{f} = \frac{1}{u} + \frac{1}{v}$$

symmetrical case gives :

u = v = 2f

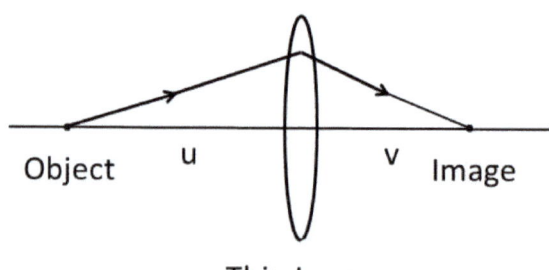

Object   u   v   Image

Thin Lens

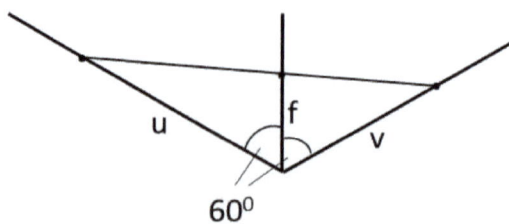

u   f   v

60^0

The same procedure can be used for resistors in
parallel, capacitances in series etc.!

Logarithmic Derivatives!

If two variables x und y are related like this:

$$y = a \cdot x^n$$

For a plot: take logarithmic scales.

For small changes:

take the derivatives with the relation

$$\frac{dy}{y} = n\frac{dx}{x}$$

$\Rightarrow$ 1% increase in x gives n% increase in y

example:

a cube with a side length L has a surface

$O = 6\,L^2$.   Its volume is $V = L^3$.

=> an increase of **1%** in the length L gives an

increase of about **2%** in the surface and an

increase of about **3%** in the volume.

For specialists: "Magical Triangle"
with logarithmic derivatives for the
trigonometric functions
s≡sinφ, c≡cosφ, t≡tanφ

$$s^2 + c^2 = 1, \quad t = \frac{s}{c}$$

$$\frac{ds}{s} = c^2 \frac{dt}{t}, \quad -\frac{dc}{c} = s^2 \frac{dt}{t}, \quad -\frac{dc}{c} = t^2 \frac{ds}{s}$$

s,c,t are on same level => **Democracy !**

This triangle can be easily remembered! Only −dc instead of +dc ist special

The factors s^2, c^2, t^2 build an inverse triangle; remember: $s^2 = c^2 \cdot t^2$

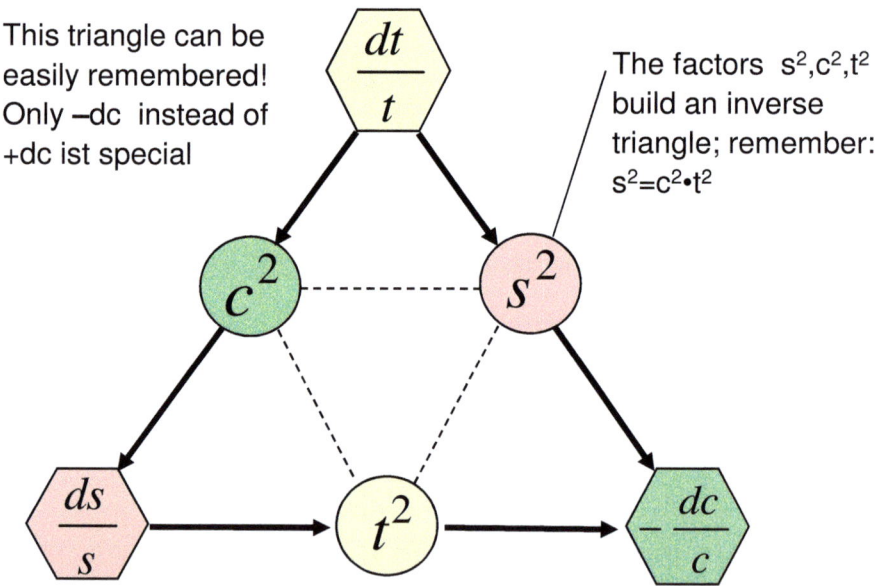

I constructed the same magic triangle for the dimensionless relativistic values for total energy γ, momentum p/mc and velocity β.

Phoning "the speed of light" to Greenland

Up to 1983 the speed of light (= c) was determined by physical measurements.

But in this year, the "Standard Meter" was given up as the standard unit for length.

Instead the **speed of light was defined** to be:

$$C = 299'792'458 \ m/s$$

I tried to phone "the speed of light" as a phone number to Greenland, since 299 is the country code for Greenland. But this number does not exist (yet).

In 2011 I wrote a letter to Mr Jens Frederiksen, responsible for the telephone services in Greenland. I suggested that he should install such a number, and that a person phoning this numer would get e.g. some information about science projects in Greenland.
I did not get an answer.
Maybe somebody else is more successful?!

In Switzerland there is no area code with 029, but maybe in some other countries?

With an interest rate of 2% it takes 35 years to double the income (50 years without compound interest)

How can we get this result very quickly?

For a quick estimate of exponential growth one uses:

$$e^7 \approx 2^{10} \approx 10^3$$

The **magic number** is thus **70 years** ($70 \approx 100 \ln 2$):

With an interest rate of p(%) it takes T_2 years to double the principal C_0.

$$T_2 = 70 \text{ years}/p(\%)$$

To increase by a factor of 1'000 ($\approx 2^{10}$) takes T_{1000} years:

$T_{1000} = 10\, T_2 = 700 \text{ years}/p(\%)$

Savings Account of William Tell

700 years ago William Tell opened a bank account with 1 frank. He then decided that the oldest child should always inherit this account. With an interest rate of p % this account is worth today 10^{3p} franks.

At **3% interest** this amounts to 10^9 **= 1 billion franks**

But 1/3 of the annual interest always went to the state in tax. Thus the net interest rate is only **2%**.

=> The sole inheritor of Tell's account owns today "only"

10^6 **= 1 Million Fr.**

Where are the remaining 999 Millions?

The state takes it all !

It is hard to believe, but the remaining **999 Million franks belong to the state!** Because the state does not pay taxes to himself. It gets the full 3% interest on his income from taxes, which it deposits in the bank as well. After one year the state owns only 0.01 frank from the income tax of Tell. But after 70 years the state owns 4 franks, the same amount as Tell. But now the difference between the exponential growth of 2% and 3% comes fully into play! After 200 years the state profits more from the 3% interest on its own property (10 franks/year) than the tax income from Tell's inheritor (0.5 frank/year). And after 300 years the state practically does not need the tax from Tell anymore!

Growth of Capital

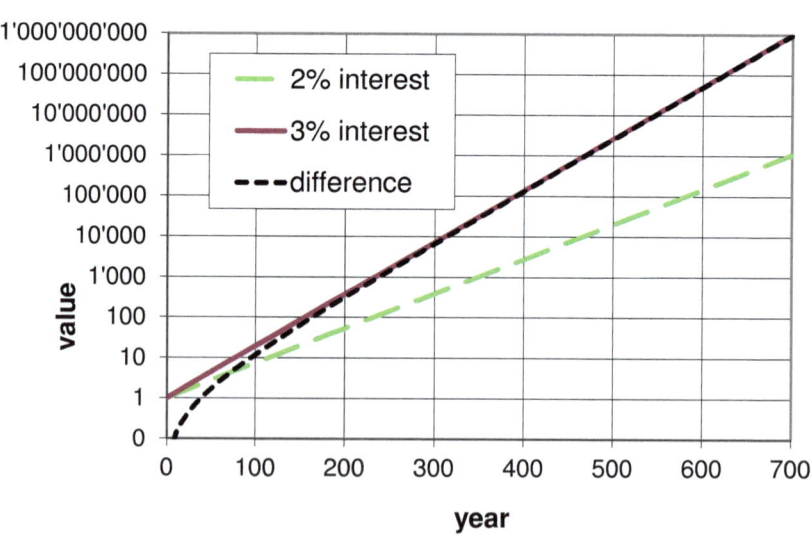

50

Stars and Galaxies in the Universe

In our galaxy, the milky way, there are approximately

400·10^9 stars.

… and in the universe there are an estimated

200·10^9 galaxies.

To put these numbers in relation with the people on Earth:

Each person on Earth can give a personal name to

50 stars in our galaxy

… and can give personal names to

25 Galaxies in our universe !!

How much longer does humanity exist?

Hypothesis: till every person who ever lived gets his own star in our galaxy!

=> until about **400 billion people have** lived on earth**.**

According to the old testament (Exodus 32; 13) God swore to Abraham, Isaac and Israel: "**I will make your descendants as numerous as the stars in the sky**".

About 100 billion people have lived so far. Let's suppose that the population grows till it reaches about 30 billion people in the year 2300, and stays constant after that. Then in the year 2800 about 400 billion people will have lived, and each of them can own a star.

=> **Humanity ends after another 800 years !**

Birth and Mortality Rate World Wide

How many babies are born each second?

And how many people die each second?

Simple model: average lifetime = **70** years

With 7'400 million people on earth the average "**exchange rate**"
for a **constant** population is 7'400 million /**70 years** or
≈ 100 million people/year. But the population is growing steadily.
The mortality rate is thus lower, and the birth rate higher than the
average exchange rate: **Mortality rate** ≈ average exchange rate
35 years ago (average age of those who die today)

= 4'600 million/70years ≈ 66 million/year

The anticipated population in 35 years (about 10'000 million
people in 2050) can be used for the present-day birth rate:

Birth rate ≈ "exchange" rate in 2050 = 10'000 million/70 years

= 140 million/year

The actual numbers in 2012 were:

≈ 136 million births/year ≈ 4.5 babies/s

≈ 56 million deaths/year ≈ 2 deaths/s (20 % from hunger!)

This amounts to **2.5 additional people on earth every second**!!
The application of this model to cities or countries is not so simple
due to migration, living habits etc. For cities in Switzerland the
approximate rule is: **mortality rate ≈ population/ 120 years**

The world population on a moving belt

In this model the world population, at the moment about 7.4 billion people, are standing on a moving belt. This belt is on the equator, encompassing the whole earth. The people form **40 million rows of 185 persons** positioned 1m apart. The distance between neighbours is 1m as well. The width of a row is thus 185m on average. The belt moves around the earth in one century. Not many of us survive long enough to make one full revolution, the majority of us is pushed over the belt into death before this period. Since the population grows steadily, the row with the newly born is about 360m wide. It takes only 80s to fill a new row, corresponding to the previously mentioned 4.5 new babies each second.

How fast is this belt moving?

The velocity is 40'000 km/100 years. This is a leisurely pace of

13 mm/s   or   **45m/h**

=> **All the people move one row forward each 80 seconds.**

To look at this in a different way: it is as if you would

move every 5 days to the next village, 6km away.

1'000 birthday balls

Today about 7.4 billion people are living on the Earth. Imagine that, before you were born, you could draw one single ball from a bowl with 1'000 balls. This ball determined where you were born. Each ball represents about 7 million people. As an example: I myself have chosen the only ball which was marked "Switzerland"!

The majority of these 1'000 balls, namely about 600, represent a country in Asia. For Africa there are 150 balls. The rest are distributed among the other continents: Europe including Russia 110, Latin America 80, North America 50, and Australia 6.

If you are unhappy with your life, would you rather have another chance?

Probably not. If you can read this article you are already privileged, because on 500 of the original balls was a note saying:
You will have no access to sanitary installations and you will not earn more than 3$ per day !
On 140 balls is written: **you will always be hungry.**

Your new gender will be chosen by a toss of a coin. There is thus a 50% chance, that it will change!

special dates in your life

You lived

1 million minutes after about 2 years

100'000 hours after about 11 years and 9 months

1'000 weeks after about 19 years and 2 months

10'000 days after about 27 years and 5 months

1 billion seconds after about 31 years and 9 months

and: if you are young,

there are worldwide about **350'000 persons**

who were born on the same day as you!

And about **250'000** if you were born around 1940.

Heart ⇔ Motor

Which is more reliable, your heart or the motor of your car?

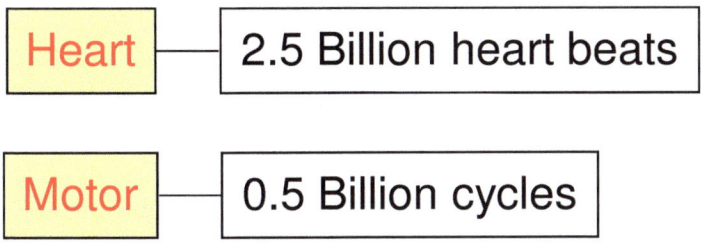

| Heart | 2.5 Billion heart beats |
| Motor | 0.5 Billion cycles |

Assumptions:

- a car can make about 200'000 km with an average speed of 50km/h
  => it runs for about 4'000 h or 240'000 min.

- the motor runs at an average of 2'000 cycles/min

=> the motor makes about **0.5 $\cdot 10^9$ cycles**

If you live 80 years, your heart has made about

2.5 $\cdot 10^9$ heart beats (non stop!)

=> your heart will make about a factor 5 more cycles than the motor of a car!!

Theorem on Salary

Dilbert's Theorem on salary states, that Engineers, Teachers, Programmers and Scientists can never earn as much as Business Executives and Sales People.

Mathematical proof (unknown source):

Postulate 1: Knowledge = Power (Knowledge is Power)

Postulate 2: Time = Money (Time is Money)

Postulate 3: Power = Work/Time (law of physics)

$\Rightarrow$ Knowledge = Work/Time (since Knowledge = Power)

$\Rightarrow$ Knowledge = Work/Money (since Time = Money)

Solving for Money we get:

Money = Work/Knowledge

=> As Knowledge approaches zero, Money approaches infinity, regardless of the amount of work you do.

Conclusion:

The Less you Know, the More you Make!

$$E = mc^2$$

The Swiss physicist Paul Scherrer gave a beautiful analogy (mentioned by Max Frisch in "Stiller") for this famous Einstein equation:

„This energy E is deposited in a blocked bank account"!

$E = mc^2$ gives an enormous energy density

- 1kg $\Leftrightarrow$ 10^{17} J = 25 TWh ,

  energy consumed in 5 weeks in Switzerland !

- Fission of U_{235} – nucleus with Neutrons

  => chain reaction

- 1 kg natural-Uranium => 7 g fissionable Isotopes U_{235} ,

  => 7 mg can be converted into kinetic energy

  = 175 MWh , equivalent to 20'000 lt Oil

- => enough energy to put a 30 t lorry

  into a satellite orbit !

Is Antimatter the Energy Solution ?

Visit of a sattelite from a galaxy with Antimatter :

=> 10 ton of Antimatter

1. with controlled annihilations:

   energy supply for the whole earth for 3 years !

2. with uncontrolled annihilations:

   energy of 30 Million Hiroshima bombs !!

A Collection of Riddles

1. A snail climbs up on a pole, 10m high. During the day it moves up 3m, at night it slips back 2m during its sleep. After how many days is the snail on the top?

2. A bottle of wine costs 11 franks including the package. The bottle is 10 franks more expensive than the package. How expensive is the package?

3. A farmer wants to fence in a square piece of land. How many posts does he need if he wants six posts on each side?

4. A water lily in a pond doubles its area each day. After 20 days, the pond is fully covered. When is the pond ¼ full with the lily?

5. There are five apples in a basket. Distribute these apples among five kids. But at the end, an apple should remain in the basket.

6. In the attic there is a box with 20 green and 20 red socks. A mother asks her son, who is colour-blind, to get a pair of socks. How many socks has the son to carry down, if mother wants
a) a pair of socks, no matter what colour?
b) a pair of red socks?

7. The stork has to bring seven babies to three different families. But no family should receive triplets. How does the stork manage this?

8. Each full hour a city train leaves Zurich for Geneva.
At the same time another one leaves Geneva for Zurich.
The time for each journey is exactly 2h 58min. How many other
city trains does each train cross?

9. A stone weighs 1kg plus one half of its weight. How heavy is it?

10. 4 hens lay 4 eggs in 4 days. How many days does one hen
need for one egg?

11. What is worth more: 1kg of 2 frank silver coins or 2kg of
1 frank silver coins?

12. The time between the first and the last ring of a church bell
is 30 seconds at 6 o'clock. What is the time interval at 9 o'clock?

13. What is the opposite of "not inside"?

14. You want to cook "spaghetti al dente". This takes 12 minutes.
But you have only two hour-glasses with 8 and 5 minutes. How do
you proceed?

15. Tomorrow a relative will visit me: his father is the father of my
father. Who is it?

16. A girl kisses a man on the street. Her girlfriend asks: "who is
this man?" The girl answers: "his mother is the mother in law
of my mother".

17. A wheel has 36 spokes. How many empty spaces does it
have in between?

18. The three sides of a triangle are 13, 31 and 18 cm.
What is its area?

19. A steamship on a river needs 9 hours from A to B.
From B to A it takes only 6 hours. How long does it take a piece of
wood to float from A to B?

20. You buy a pair of shoes with a discount of 10%. What is
cheaper for you: first to subtract the discount and then add the
8% sales tax, or first add the tax and then subtract the discount?

21. Two girls are born on the same day from the same parents.
But they are not twins. How come?

22. I walk 5km to the south, then 5km to the west and finally
5km north. I find myself back at my starting point. Please explain.

23. BSAINXLEATNTEARS.
Cross out six letters to get a familiar word.

24. In my purse I have two bills with a total value of 110€.
But one of the bills is not a 10€ bill. How is this possible?

25. During one day a share of a company loses 30% of its value.
The next day the value of the share increases again by 30%.
So everything is back to normal again, right?

26. How often does the number 8 appear in the numbers from
1 to 100?

27. How often do the big and the small hands of a watch overlap during 24 hours?

28. John bought 13 chestnuts and ate all of them except four. How many does he still have?

29. How many months have 30 days?

Now it gets more ambitious!

30. A man walks to his office each morning. In the evening he jogs back home. For both journeys he needs 35minutes in total. If he walks both ways he needs 50 minutes. How long does he need to jog home?

31. After work a man takes the train home and is picked up by his wife at 18.00 at the railway staion. One day he takes the earlier train and is at the station already at 17.00. He decides to walk towards his wife and is thus 70 minutes earlier at home. How long does he walk before meeting his wife?

32. Maria is 24 years old. She is now twice as old as Anna was, when Maria had the present age of Anna. How old is Anna now?

33. In order to finish 1km of highway 20 workmen need 60 days. How long does it take 12 workmen to finish 500m?

34. A train travels at 80km/h from Zurich to Geneva. At the same time a train leaves Geneva for Zurich, together with a dove. This train has a speed of 70km/h, the dove flies at 100km/h. As soon as the dove meets the train from Zurich, it flies back towards the train from Geneva. It oscillates between the two trains, till they meet. How far does the dove fly, if the distance from Zurich to Geneva is 300km?

35. John needs 8 hours to ride his bike from Basel to Lugano. His friend Max lives in Lugano and needs 12 hours from there to Basel. Both start at the same time. When do they meet?

36. 152 players want to participate in the tennis tournament in Wimbledon. But in the first round there is room for only 128 . Of these,120 qualify automatically, while the other 32 players have to play two qualifying rounds in a cup system to determine the remaining 8 players. How many games are required alltogether to determine the winner?

37. It is snowing continuously. A snow plough starts on a long road at 10.00. In 1 hour it cleans 1km. During the next hour only 0.5km. When did it start snowing?

38. Three friends take a drink at a bar and pay 10 franks each. Then the owner of the bar enters and sees his his old friends. He gives the barman five franks with the order to give it back to his friends. The barman cheats and gives every guest only one frank back. Every friend has thus effectively paid nine franks, altogether 27 franks. The barman keeps two franks. Question: where is the 30th Fr.?

39. The two school buddies Jim and Henry meet again after a long time. The following dialogue develops:
- Do you have children?
- Yes, I have three.
- How old are they?
- Multiplying there ages together gives 36.
- Well, this does not help me much.
- OK, adding their ages gives my house number here.
- Sorry, this is still not enough information for me.
- I will give you one last hint: the oldest son loves to play football!
- Thanks Henry, now I know the ages of your children!

40. It takes six minutes to fill a bath tub, and ten to empty it. How long does it take to fill the bath tub, if the drain is open while the water is let in?

41. In a quiz a pot of coins can be won. All coins have the same value. Ten people are participating. Those who answer three questions correcly share the coins. How many coins have to be in the pot, so that for all possible outcomes each winner gets the same number of coins?

42. A sheikh has three sons. In his last will he states, that the oldest son will get half of his camels, the middle son 1/3 and the youngest son 1/9 . When the sheikh dies, his fortune consists of 17 camels. How do the brothers proceed?

43. Peter told me: the day before yesterday I was 29 years old. Next year I will be 32 years old. When did Peter tell me this, and when is his birthday?

44. A sheikh has n camels and sells them at n dinars each. He gets the total sum in 10-dinar bills plus a small amount in coins. He distributes this evenly to his two sons. At the end a 10-dinar bill plus the coins remain. He gives the bill to the elder son and the coins to the younger son. Afterwards the younger son complains that this is not fair. The older brother responds with a gesture: OK, he says, to even things up I give you my pen. What is the value of his pen?

45. Max tells his friend: my father is five times older than me. But in three years he will be only four times older. How old is Max?

46. Each Tuesday the members of a hunting club meet at a round table and tell their hunting stories. There are two kind of hunters: those who always tell the truth…and those who always lie. One day a hunter says: Aha, today I am sitting between two liars! Then everybody looks at his two neighbours and exclaims: That's funny, that is the case for me as well!
At the next meeting the vice-president says: what a strange coincidence at our last meeting, when we had twelve people around the table! The cashier of the club responds: no, we were really thirteen! Who is telling the truth and who is the liar?

47. Did you fall into the trap of riddle 19? If not, I present here the alternate riddle which represents a real situation and requires some algebra to solve it:
A steamship on a river needs 9 hours from A to B.
From B to A it takes only 6 hours. How long does it take a piece of wood to float from **B to A**?

Solutions

1. After 7 days the snail slips back from 9m to 7m.
   On the 8th day it makes it from 7m up to the top.
2. 0.5 Fr.
3. $20 = 4 \times (6-1)$; the corner posts can be seen from two sides, thus count only once.
4. After 18 days
5. Give the 5th kid the basket with the apple in it.
6. a) 3; b) 22
7. What? You still believe the stork story?
8. 5; each train meets another one every 30 minutes.
9. 2kg
10. 4 days.
11. 2kg of silver are worth twice as much as 1kg silver.
12. 48 seconds (6 seconds between 2 rings)
13. inside
14. Both hour-glasses start simultaneously. After 5 minutes the small one is empty and is turned over. After 8 minutes the big hone is empty as well and is turned over. After 10 minutes the small one is empty again. In the big hour-glass there is sand for 2 minutes at the bottom. If it is turned over, then after 2 minutes it is empty and the desired 12 minutes are over.
15. The uncle
16. The father or the uncle
17. 36 as well
18. zero! $18+13 = 31$; this means we have a straight line.
19. The wood can't float against the stream, which is from B to A.
20. The sequence does not matter! If $F_1 = 1$-discount;
    $F_2 = 1$+sales tax; Price $= F_1 \times F_2 = F_2 \times F_1$
21. They are a part of triplets or quadruplets

22. There are two solutions. The first one: start at the north pole. The second solution: close to the south pole there is a circle of latitude of length 5km. Just start 5km north of it.

23. BANANA; you have to cross out the letters SIXLETTERS

24. One of the bill is 100€, the other one 10€.

25. The share increases 30% on a value which decreased to 70% the day before. Its new value is now only 91% (1.3x70%). Only with an increase of 43% do you get back the original value.

26. 20 times. If you write all 100 numbers from 0 to 99 with 2 digits as 00, 01, 02,…98, 99, you need 200 digits. Every digit from 0 to 9 appears equally often and thus 20 times.

27. 22 times. The big hand makes 24 revolutions per day, the small one 2 revolutions; the difference is 22 (like in a race).

28. Exactly those four.

29. All months except February.

30. 2x(walking + jogging) = 70 minutes; 2x(walking) = 50 minutes => 2x(jogging)= 20 minutes; => jogging = 10 minutes.

31. 25 minutes. The 70 minutes the man saves are the time his wife would need from the meeting point to the station (=35 minutes) and back (another 35 minutes). Thus the meeting occurs at 18.00 - 35 minutes = 17.25, 25 minutes after the man arrived.

32. Anna is 18 years old = (24+12)/2; with some algebra: today: Maria=M=24; Anna=A, x years ago: $M'\equiv M-x=A$; $A'\equiv A-x=0.5 \cdot M$; => $M-A=A-0.5 \cdot M$; $4A=3M$, $A=3/4 \cdot M=18$ years. x=6 years.

33. 50 days (=0.5x60x20/12)

34. The two trains approach each other at a combined speed of 150km/h. Since the two cities are 300km apart, they meet after 2 hours. Thus at a speed of 100km/h the dove flew 200km.

35. After T=4h48min. (1/T=1/T1+1/T2)

36. After each game the loser drops out. Every player except the final winner loses exactly once. Thus we have 151 games, always one less than the number of participants.
No calculation is needed!

37. At 9.30h. The snowfall is constant. At 10.30 the height of the snow is exactly the average of the hour from 10 to 11h, (lets say 10mm). At 11.30h the height is the average of the hour between 11 and 12h and thus twice as high (0.5 km against 1 km covered) as the average of the previous hour (20mm). Calculating back, the snowfall had to start at 9.30h.

38. The 2 franks have to be subtracted, not added!
27 − 2 = 25 franks (and not 27 + 2 = 29 franks)

39. There are eight ways to get a product of 36 with three numbers. Adding the corresponding numbers give 8 possible house numbers: 10, 11, 13, 13, 14, 16, 21, 38. Since the house number is not unique to determine the ages, it has to be the number 13, with the two possibilities 1, 6, 6 and **2, 2, 9**. Only in the second case do we have an oldest son.

40. 15 minutes, because in 30 minutes the bathtub would be filled five times and emptied three times.

41. n, the number of coins has to be divisible by all numbers from 2 to10. => n=10x9x7x4= 2'520. With 13,14, 15 persons, n increases dramatically to n=360'360.

42. They borrow one camel from a neighbour. With 18 camels the oldest son gets 9, the middle on gets 6 and the youngest son gets 2 camels. Together 17 camels get distributed and they can give the 18th camel back to the neighbour!

43. On the 1st January of this year Peter told me that yesterday, on December 31st, he had his 30st birthday. He was thus still 29 years old on the 30st December of last year. This year he will be 31 on December 31st. Next year on December 31st he will be 32.

44. Subtract from all squares n^2 a multiple of 20 (corresponding to the two bills of 10 dinars). The remainder (= n^2 modulo 20) is thus between 0 and 19. In order to have a 10-dinar bill left together with some coins, the remainder has to be greater than 10. But only with 4 and 6 as the last digits of n^2 is the remainder larger than 10 (=16 in both cases). From these 16 dinars the older son gets 10 dinar as a bill, the younger son 6 dinars in coins. To even things up the **pen** has to cost **2 dinars**.
45. Max is 9 years old.
46. It had to be 12 hunters. Lets give all liars uneven numbers, and even numbers to the others. With 13 hunters, liar 13 would sit next to liar 1. This leads to a paradox.
I sent this riddle to Martin Gardner, who published it in November 1967, Vol. 217 in his "Scientific American" column, and in a book which was published in German as "mathematische Hexereien" (page 65).
47. The wood floats T=36 hours on the river. If T1= 9 hours, T2=6 hours, then $T=2 \times T1 \times T2/(T1-T2)=36$ hours.
The velocity of the ship on a lake would be 5 times, = $(T1+T2)/(T1-T2)$, the velocity of the river.

Remark:

I collected these riddles over many years from different sources, which I can no longer identify, I have to admit.
Most of them have been known for a long time. But some of them come, to my knowledge, from myself; their numbers are: 2, 3, 10, 18, 20, 25, 36, 38, 46.

No risk, no fun !

The author enjoys some "risky walking" in the
Alpine foothills of Appenzell in Switzerland (2009)